Where is Quality Control in Commercial Construction?

Project –"A"

The cover photo is not a mock up. It is not a joke, but instead it demonstrates firsthand an example of how serious the issue of Quality Control in construction has become. This beam pocket was designed originally to have an embedded plate with anchors. The contractor missed not only placing the embed but creating the pocket altogether.

The contractor approached the design team with a request to cut out a beam pocket and use adhesive type anchors to install the beam bearing plate. First, they cut the pocket too deep, requiring several shims to bring the beam to the correct elevation. Secondly, their attempt to install the adhesive anchors was far from acceptable.

If this was the only issue on the project it would have been one thing. But the reality is that this single story project had so many quality control issues that the original schedule of 9 months for construction took over 2 years to complete. All of the issues were directly tied to a lack of supervision by qualified personnel on by the G.C.

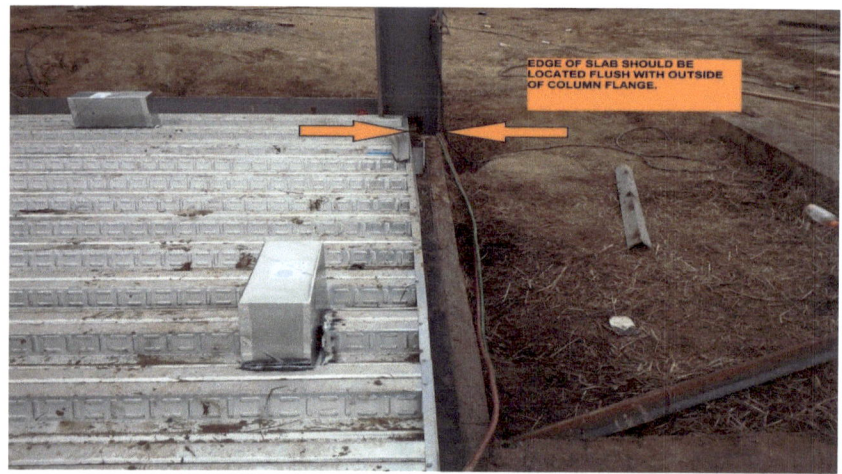

Another issue on the same project, this was clearly shown on the construction documents but if you can't read plans I suppose there would be some confusion.

Note the cap plate on this column does not match the slope of the beam. While this may be an easy fix, it shows that quality control goes beyond the jobsite and into the fab shop

Concrete columns were laid out in the wrong location, all had to be removed and new columns formed & poured in the correct locations creating more job delays.

Several anchors were added to column locations on site, drilled and set with adhesive, all of the additional anchors were set too low and coupling nuts had to be used.

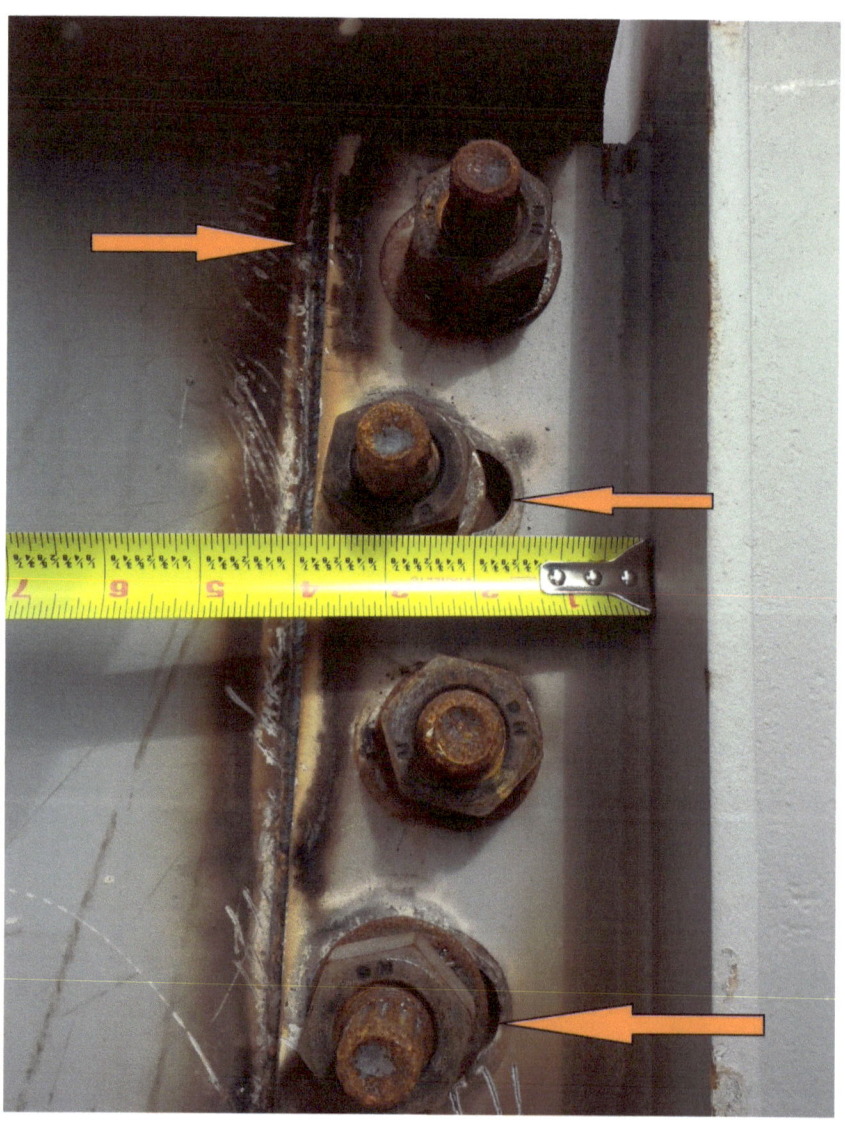

Another example of what lack of supervision in the field can result in.

Vapor barrier should have run all the way to the edge of the new concrete slab. Wire mesh had to be removed to access the area needing additional vapor barrier coverage.

Several other locations on the project where complete vapor barrier was lacking.

Quality control continued to be an issue as new concrete columns were being formed. Clearances required for reinforcing bars were not met.

Several reinforcing bars were noted as not even tied to adjacent bars

Project –"B"

Grouting of baseplates serves no useful purpose if not done properly

This contractor applied the grout thru an open hole in the base plate while placing forms around the perimeter. No way for him to see if there was settlement until later.

In this case, tension control bolts cannot be tightened as they are blocked by the plate creating the joist bearing seat. What good are connections if bolts are not tightened?

And last but not least, a seismic brace frame with no weld on the top connection. Normally this would not be alarming, just considered work in progress. However, since all other brace frames on the project were completely welded on the 3 floors above, I assume this is just another case of someone who walked off without completing the job since it was lunch time, or 5 o'clock.

Actually, Project B was one of the better projects inspected this year…

Project – "C"

Main electrical panels being used on the jobsite for a plan rack…..seriously?
This photo is not staged, it simply shows how the subcontractor doesn't care about the
quality of his work, and how supervision by the G.C is seriously lacking.

Brace frame column w/ 3″ thick base plate and 2.5″ dia. anchors which are too short

Seismic loads for this column require all anchors to catch the full nut to develop the connection. Coupling nuts had to be obtained to save this column connection.

Anchors too short and off pattern. Subcontractor torched the base plate without consulting the design team, and cut the plate within ½″ of the edge. He then welded plate washers over top of the oversize hole to cover it from inspection.

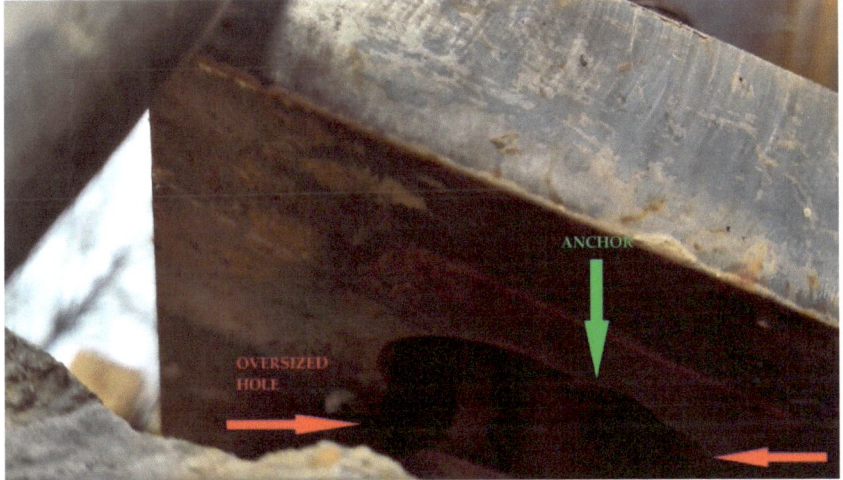

The anchor in this photo is ¾″ the oversize hole is nearly 3″ in each direction.

Project –"D"

Classic example of no one paying attention to the subcontractor. The laundry room is set up for 4 dryers. For this to work you need 4-electrical outlets and 4 vents, not 3.

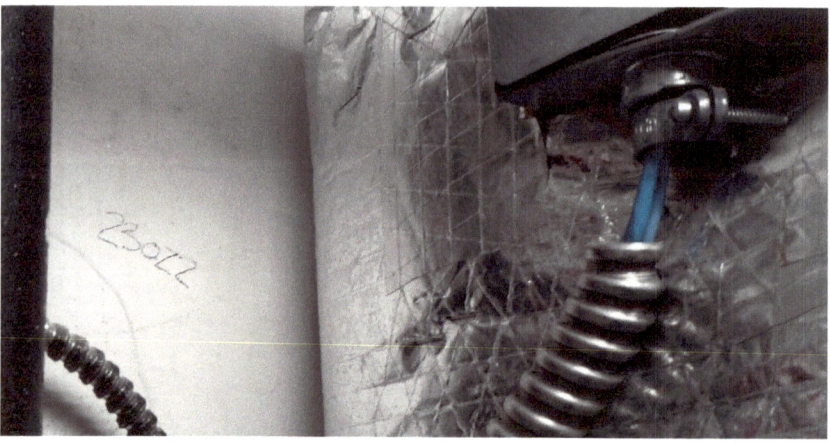

Throughout Project "D" the electrical subcontractor continued to make this mistake repeatedly, and no one addressed it until the design team did their punch list.

Exit signs that are properly installed will not fall out of the ceiling ......it's a proven fact

Overflow switch should be placed inside the drip pan of the air handler so that it can detect water and shut down the unit if necessary. In this case, the cleaning crew came thru to clean the drip pan and left all the sensors on the outside of the pans. The GC didn't catch any of these, and no one supervised the cleaning crew.

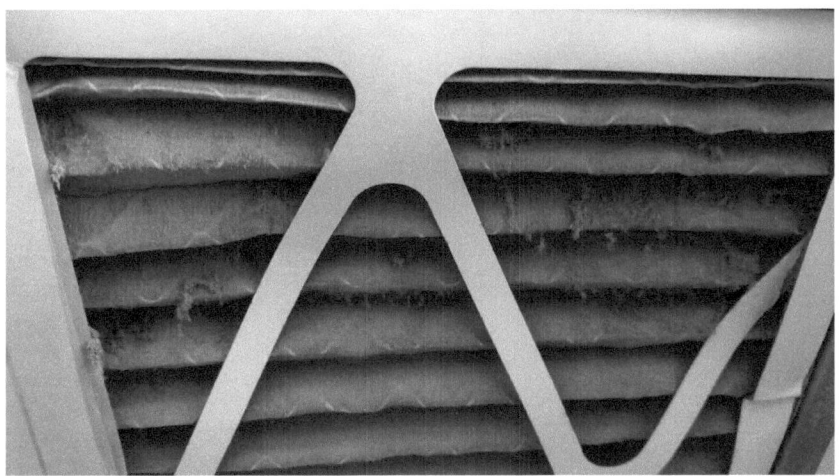

Filters in the units are to be changed regularly during construction and especially when turning the building over to the new owner. The GC didn't check to see if the sub had completed this task, and most all of the filters looked like this one or worse.

This is a simple case of the installer not understanding what fire caulking actually does and the superintendent not coming behind him to check on the quality of his work.

The Case of the Mysterious White Dust

Sometime early in March 2014., I received a call from an Architect with concerns over a mysterious white dust that was coming from the air conditioning supply vents on a project we had worked together on the previous fall. Our office had provided the mechanical design for the new school addition, and claims were being made that the white dust coming from the air vents must be a design issue. This seemed very strange since the units were composed of simple air handlers & outside condensor units.

Nevertheless, I made the trip down to the project to meet the Architect, Owner & Contractor to see if we could determine what the issue was. Our findings proved very interesting as you will see in the report on the following pages;

1. Walk through of the project was conducted with Architect, General Contractor, Mechanical Sub & School personnel to investigate the source of white dust in the mechanical systems.

Initial Comments before going in:
Parking Lot – Discussion with the contractor regarding the doors of the facility were open during the schools moving in furniture. Architect stated that certain doors were noted as closed during parking lot work, and that the building still was in contractors control and had not been turned over to the owner at that time.

Inside the building
A random mechanical closet was opened to take a close look at one of the units. The mechanical subcontractor removed the covers from the unit exposing the coils and fan wheel. Subcontractor noted that the fan wheel had lots of dust build up and the coils were full of dust also. **(See Photo #1)** The other thing noted is that the filter had not been fully inserted into position, leaving approximately 2" of the return unprotected and allowing unfiltered air to enter into the unit. **(See Photo** #2**)** This seemed very odd.

Second Mechanical closet was opened and in this case, the filter again was not inserted completely back into place to provide protection to the air handler, leaving approx. 4" of unprotected space in the return for unfiltered air to enter into the system.**(See Photo #3)**

In every case where the filter was not properly installed, it was noted that a good portion of the return air was being pulled through the door frame instead of through the return air grille, and this being the case, there was excessive dust around each of these door frames. (**See photo #4**). In one case where the filter was properly installed, there was no dust evident around the door frame indicating all air was going through the return.

It was noted that the most of the filters were much in need of being replaced at this time.
Maintenance staff stated they had been changing filters every two months.
(See Attached Photos)

Coils in the first unit looked at were full of dust as was the fan wheel. This appeared to be caused from the filter not being properly installed allowing unfiltered air into the system.

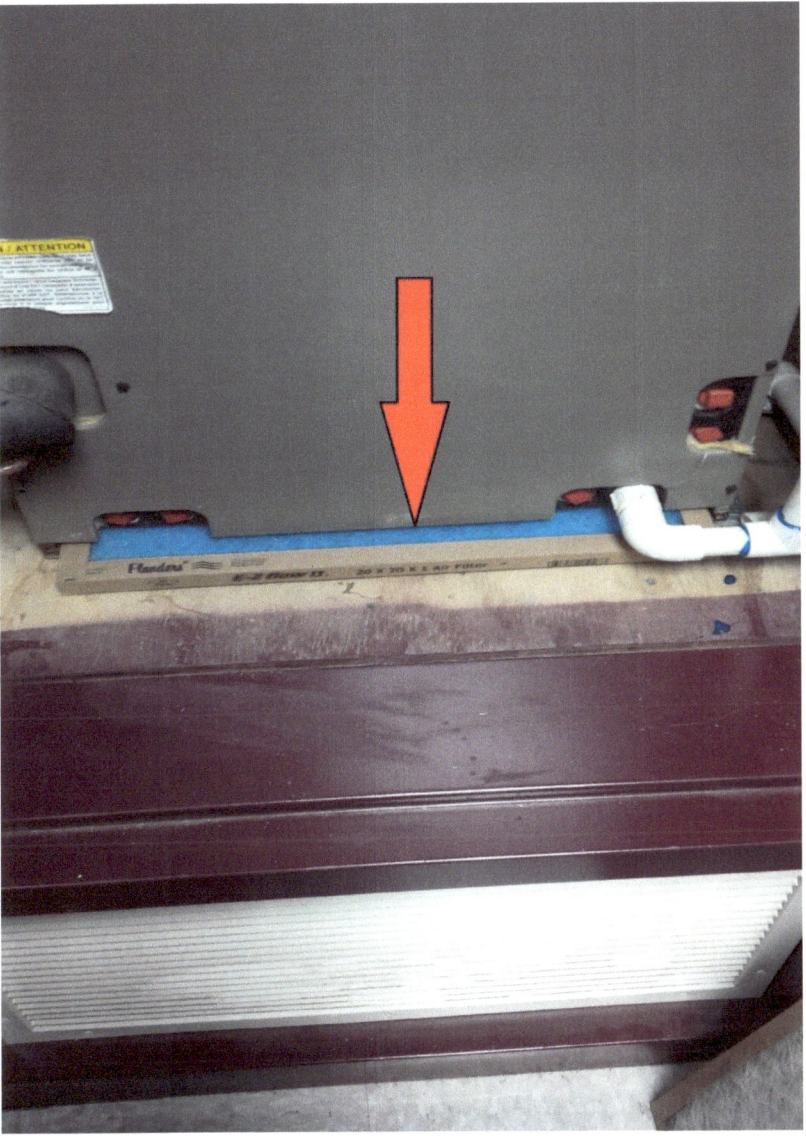

Two issues, Filter is not pushed back fully into place allowing unfiltered air to enter through the back of the unit, and filter cover panel is not in place –allowing air infiltration at front of the unit

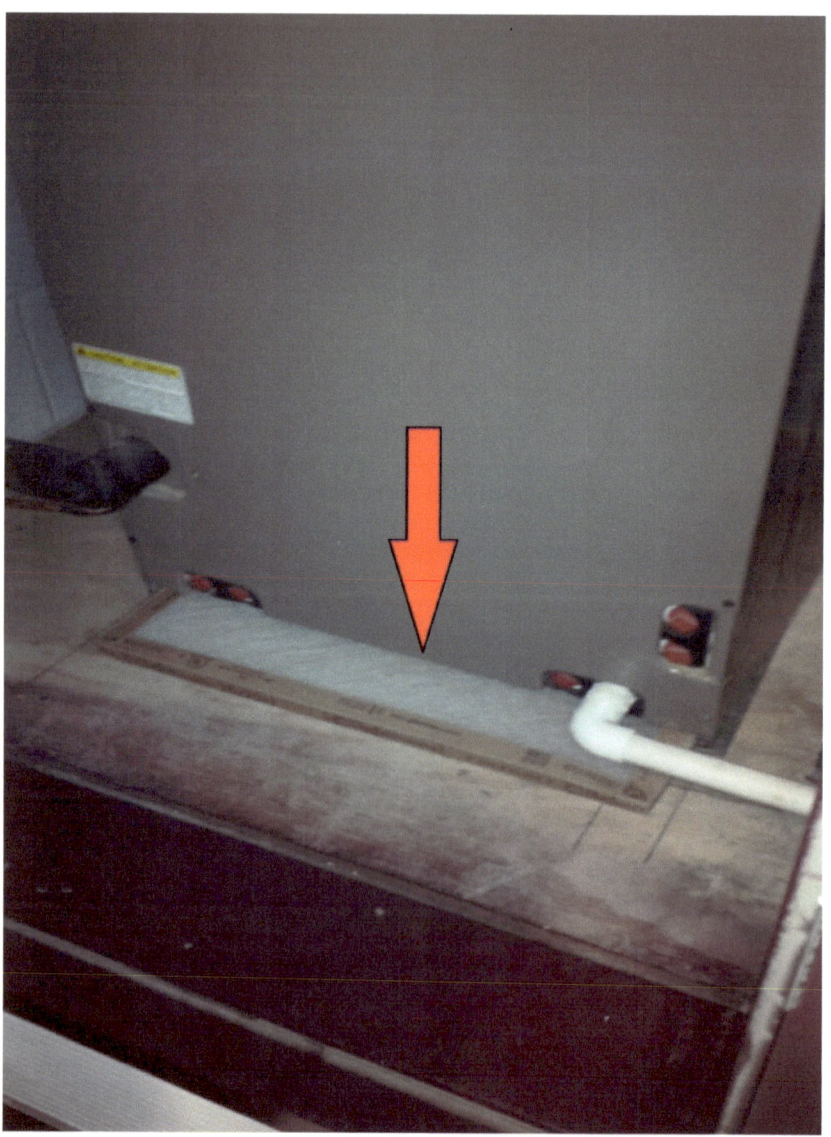

Filter not properly installed allowing air to pass into the system unfiltered, note that due to the filters position, the filter cover panel is not installed either

In each case where filters were not installed properly, there was excessive dust around the door frame. Clean area on frame is where we wiped it down. Where filters were properly installed, there was not any dust around the door frame.

General Items;
In a discussion with the maintenance staff we discovered the following timeline;

August 2013 – Contractor turns facility over to school with pleated filters in all units
October 2013 – (2 months) Maint staff changes filters with other type filters
December 2013 - dust begins to appear two months after the other type filters are installed

It appears to have taken about two months after the filters were replaced for dust to begin entering into the system and build up to the point where it became visibly evident.

Final Note;
At first sight this issue seemed to be the result of the schools maintenance staff not properly installing the new filters. With the new filters inserted only ¾ of the way in, unfiltered air was entering the units and dust was accumulating. Upon further investigation it was later discovered that the reason the filters were so difficult, nearly impossible to insert fully into the unit was due to the mechanical contractors work during setting of the air handler.
When the mechanical subcontractor cut the hole in the mechanical platform for the return air to enter the bottom of the unit, he didn't set his saw blade to the depth of the plywood, but instead cut the hole in the platform using his sawblade at full depth which cut almost completely through the supporting structure. Once the air handler was installed on top of the platform, the framing gave way to the weight of the unit creating distortion in the track which was to receive the filter. Most units it was impossible to reinsert the filter more than ¾ depth.

Once again, lack of supervision, lack of skilled labor resulted in substandard construction causing time and expense to the Owner and the Designers.

The photos on the previous pages reflect just a few of the issues which I have personally encountered on commercial construction projects in the past year. The information in this brief publication is not meant to downgrade contractors or sub trades in general. There are many tradesmen out on the job who do it right and care about quality. However, there seems to be an increase in the lack of supervision on commercial projects, and with time constraints along with tight budgets – the issue just seems to be getting worse. Some of the items can be attributed to having a supervisor on the job who just doesn't have the years of experience it takes to catch all the things that come up on the job.

Hopefully this will enlighten the reader to what is going on each day out in the field, and hopefully drawing attention to the issues will make all in a position of responsibility take their job a bit more seriously……..to the benefit of all.

Other Publications by the Author – Available on Amazon Kindle E-Books

-Basic Guide to Project Management in Commercial Construction;

-Step by Step Guide to Home Construction

Classic Kitchen Renovation – from 1970 to 2016